ACKNOWLEDGEMENTS

I would like to thank all the sales representatives and managers I have worked with over the years for showing me the difficulties of putting theory into practice in the field – and my husband and daughter for tolerating my lack of putting theory into practice in the kitchen while I wrote this book.

All rights reserved. No part of this publication may be reproduced, stored in a retrieval system, or transmitted, in any form or by any means, electronic, mechanical, photocopying, recording or otherwise, without the prior permission of J. Barnard Publishing.

This book is sold subject to the condition that it shall not, by way of trade or otherwise, be lent, re-sold, hired out or otherwise circulated without the Publisher's prior consent in any form of binding or cover other than that in which it is published and without a similar condition including this condition being imposed on the subsequent purchaser.

Although every precaution has been taken to ensure the accuracy of this publication, the author and publishers accept no liability for errors or omissions. The author and the publishers shall not be liable to any person whatsoever for any damages, by reason of any mis-statement, error or omission, negligent or otherwise obtained in this book.

First published 2001

© J. Barnard Publishing 2001

ISBN 1 872711 02 2

Published by J. Barnard Publishing, 5 Bramalea Close, London N6 4QD
Tel: 0208 341 2963 Fax: 0208 341 3018
Printed by Ink Design and Print Ltd, Bourne End, Bucks

CONTENTS

WHY SHOULD I READ THIS BOOK? — 7

CHAPTER I – THE CODE

WHAT IS THE CODE?	9
WHAT DOES THE CODE COVER?	10
HOW DOES A COMPANY KEEP WITHIN THE CODE?	12
HOW IS THE CODE ENFORCED?	13
HOW DOES THE CODE AFFECT ME?	14

CHAPTER II – THE MATERIAL

WHAT CAN PROMOTIONAL MATERIAL CONTAIN?	15
WHAT IS PRESCRIBING INFORMATION?	19
HOW CAN PROMOTIONAL MATERIAL BE USED?	20
CAN I PRODUCE MY OWN MATERIAL?	21
CAN I WRITE A LETTER?	22
CAN I SPONSOR MATERIAL?	23
CAN I USE REPRINTS?	24

CHAPTER III – THE CALL

WHO CAN I CALL ON?	25
WHEN CAN I CALL?	26
HOW OFTEN CAN I CALL?	26
HOW CAN I ARRANGE A CALL?	27

CHAPTER IV – THE DETAIL

HOW CAN I DETAIL?	29
WHAT CAN I DETAIL?	29
WHAT CAN I SAY?	30
WHAT DO I NEED?	32

CHAPTER V – THE GIFT

WHAT GIVEAWAYS CAN I USE?	33
CAN I BE INVOLVED IN A COMPETITION?	34
CAN I PROVIDE MEDICAL AND EDUCATIONAL GOODS & SERVICES?	35
CAN I SPONSOR THIS?	37
CAN I MAKE A DONATION?	38
CAN I OFFER A PACKAGE DEAL?	39
CAN I OFFER A DISCOUNT?	39
CAN I GIVE SAMPLES?	40

CHAPTER VI – THE MEETING

WHAT SORT OF MEETING CAN I HOLD?	41
WHO CAN ATTEND A MEETING?	42
WHERE CAN I HOLD A MEETING?	43
WHAT HOSPITALITY CAN I OFFER?	44
WHAT ACTIVITIES CAN I OFFER?	45
WHAT MEETING MATERIALS CAN I USE?	46
CAN I HAVE A STAND AT A MEETING?	47
CAN I MAKE A PRESENTATION AT A MEETING?	48
CAN I INVITE A SPEAKER TO MAKE A PRESENTATION?	49
WHAT CAN I PAY FOR AT A MEETING?	50
CAN I SPONSOR A MEETING?	51
INTERNATIONAL MEETINGS	52

CHAPTER VII – THE ESSENTIALS

TRAINING	53
ABPI REPRESENTATIVES' EXAMINATION	53
MEDICAL INFORMATION	54
ADVERSE EVENTS	55
EXPENSES	56

GLOSSARY 57

INDEX 59

CHECKLIST FOR MEETINGS 64

WHY SHOULD I READ THIS BOOK?

This book will help you do your job, and it will help you keep your job. It is:

- A practical and straightforward guide to the ABPI Code of Practice as it relates to all aspects of your work as a medical representative.
- Written in clearly defined sections, with key points highlighted in each section.

Detailed explanations are also given, to help you understand the reasons why you must not do something or must do something in a particular way. **Knowing the rules of the Code allows you to work effectively within them,** rather than being limited by them.

It will help you keep your job, because in many companies, breaking the Code is a disciplinary offence, which can result in dismissal. This reflects the importance both of you and of the Code.

You are important as you are, in most cases, the only direct contact between your company and health professionals.

The Code is important to your company, and to the Pharmaceutical Industry in general, because, as a self-regulatory system, it needs to be, and be seen to be, effective in order to be allowed to continue.

You should read this book if you are:

> **A new representative** – I suggest you would benefit from reading through the book from beginning to end, to make sure you are aware of all areas affected by the Code. It is useful preparation for the ABPI Representatives' Examination.
>
> **A more experienced representative** – the book can be used as a reference or reminder for specific information, particularly if your job changes significantly e.g. from a GP-based job to a job dealing with NHS management.

You should use this book in conjunction with (not as a substitute for):

> **the ABPI Code of Practice**
> **the training you receive from your company**
> **your company policy as regards Code issues**
> **experienced advice** – slight differences in situations can mean major differences as far as the Code is concerned. If you are not sure, **ASK.**

Joan Barnard
December 2000

Terminology

'medical representative' – this is intended to cover all personnel whose function includes promoting medicines directly to healthcare professionals or administrative staff.

'he' is used throughout for reasons of readability only, to avoid the more cumbersome 'he/she'.

The ***Glossary*** covers acronyms and terms used which may be unfamiliar to new representatives.

Company policy

Throughout the book, space is provided for you to record details of your company's policy in relation to specific Code issues.

CHAPTER I – THE CODE

WHAT IS THE CODE?

The full title is the ABPI Code of Practice for the Pharmaceutical Industry.

It has been in use since 1958 and is regularly revised, usually every two years, in consultation with the BMA, the RPSGB and the MCA.

The Code is voluntary. All companies who become members of the ABPI must abide by the Code and non-members may agree to abide by it. Almost all pharmaceutical companies in the UK accept the Code.

The Code is self-regulatory, administered by the PMCPA, which was established by the ABPI to operate the Code at arm's length from the ABPI itself. The PMCPA is appointed by the Board of Management of the ABPI. The Code incorporates UK requirements and the European Community Directive on the advertising of medicines. In many cases therefore, breaking the Code will mean breaking the law.

The aim of the Code is to ensure that the promotion of medicines is:

Responsible
Ethical
Professional

WHAT DOES THE CODE COVER?

The Code covers all promotion of medicines in the UK

The Code covers the promotion of all prescription medicines. OTC medicines are not covered if they are to be bought by patients, but are covered if they are to be prescribed by doctors.

Promotion is:

> 'any activity undertaken by a pharmaceutical company or with its authority which promotes the prescription, supply, sale or administration of its medicines'

This very wide definition of promotion means that many materials produced and activities undertaken by a pharmaceutical company come within the scope of the Code. Whether or not something is considered 'promotion' depends not just on what it is but also on how it is used, why it is used and what effect it has.

The Code covers every aspect of your work as a medical representative

Because your job is to sell your company's product, everything you do to achieve that counts as 'promotion'. The Code therefore covers everything that you:

Do
Say
Use
Pay for

in the course of your work.

Your involvement can mean that a non-promotional activity becomes promotional and subject to the Code.

This is why you must always follow instructions exactly – if you don't, you may breach the Code.

Certain activities are not covered by the Code

Trade practices which relate solely to financial issues e.g. discounts, prices, margins are not covered by the Code, but associated promotional activities e.g. detailing, meetings, are covered.

The Code covers the activities of third parties

Your company is responsible for the activities of any person or group acting on its behalf and with its authority e.g. meeting organisers, speakers.

If you are a contract representative, the company to whom you are contracted is responsible for you under the Code.

The Code does not cover health professionals

Doctors are governed by the GMC and pharmacists by the RPSGB. The guidance of these bodies incorporates legal requirements. *(When a health professional acts on your behalf, this activity is subject to the Code and your company is held responsible – see above under third parties).*

HOW DOES A COMPANY KEEP WITHIN THE CODE?

Promotional material is approved

Companies have different procedures for reviewing material to make sure that it complies with the Code. The approval procedure usually involves a series of reviews by different functions e.g. Regulatory, Medical Information, Medical Affairs. The final step in all approval procedures is certification.

Certification is done by two signatories, usually senior members of your company, whose names are notified to the PMCPA and the MCA. One signatory must be medically qualified. Signatories must certify that the material complies with all requirements and by doing so, they accept personal responsibility.

Certification must be complete before any piece of material can be used promotionally.

You should be aware of your company's procedure for approval and certification of promotional material.

Promotional activities are approved

Activities may be approved on an individual basis or by means of a company policy which details exactly how a particular activity should be carried out. You should be aware of your company's procedures for approving promotional activities.

All relevant staff are trained on the Code

Companies make sure that all staff who are in involved in promotion or who approve promotion understand the requirements of the Code in relation to their work (see Training p53).

HOW IS THE CODE ENFORCED?

The Code is enforced by a complaints system

Complaints are received by the PMCPA from health professionals, the MCA, pharmaceutical companies and others. The PMCPA will also pick up any adverse comment in the medical or lay press as a complaint.

A company must respond to any complaint in writing within 10 working days.

The Code of Practice Panel considers the complaint and the response and makes a ruling as to whether or not the Code has been breached.

Any decision of the Panel can be appealed, in writing and in person, to the Code of Practice Appeal Board, whose members are both independent and industry, medical and non-medical. The Appeal Board's decision is final.

Companies are penalised for breaching the Code

A company must sign an undertaking that the breach will not occur again. This may mean that promotional material has to be withdrawn from use, often at considerable cost. Failure to comply fully with an undertaking is a serious breach of the Code. It is therefore essential that you return or destroy promotional material if instructed to do so.

A company which is found to have breached the Code must also pay an administrative charge for each breach – £1000 for a Panel decision and £4000 for an Appeal Board decision. (If a company is found not to have breached the Code, and the complaint came from another pharmaceutical company, the complainant company must pay the administrative charge).

In serious cases, a company may be reported to the Board of Management of the ABPI, who may issue a public reprimand, order an audit of the company's procedures, instruct the company to publish a corrective statement or expel the company from the ABPI.

Details of complaints are published

Details of completed cases are published quarterly by the PMCPA in the Code of Practice Review which is made widely and freely available to the industry and to the medical press. The reports identify companies but do not identify individuals by name.

HOW DOES THE CODE AFFECT ME?

The Code covers every aspect of your work

Your job as a medical representative is to sell your company's product. Everything you do to achieve that counts as 'promotion'. The Code therefore covers everything that you:

Do
Say
Use
Pay for

in the course of your work.

You must follow the Code in every aspect of your work

You must follow instructions from your company about how to use promotional material (see p20). This is particularly important in relation to material which has been found in breach – make sure you return or destroy all copies exactly as instructed.

For many of the promotional activities in which you are involved, your company will have a policy which sets out exactly what you can and cannot do e.g. for expenses, for writing letters, for meetings, and you should always follow this.

You should not become involved in any activity which is not covered by a policy without making very sure that it complies with the Code.

Your company is responsible for every aspect of your work

If there is a complaint about any of your activities, your company will need to investigate this with you before responding to the PMCPA. You will almost certainly be interviewed by your manager, and perhaps by more senior members of your company. If relevant, your call sheets, expense forms and travel log will be examined and may need to be submitted to the PMCPA. Keeping accurate call reports, detailing what was discussed and what happened in each call is not only good sales practice but will also help you respond to a complaint.

Breaching the Code is a serious matter

The Code holds your company, not you personally, responsible for any breach of the Code.

However your company may hold you personally responsible. In many companies, it is a disciplinary offence for a representative to breach the Code. You could therefore lose your job.

CHAPTER II – THE MATERIAL

The following section is provided to help you understand the many different issues which have to be considered before promotional material is issued to you for use. It is NOT intended that you use this to produce material yourself – see "CAN I PRODUCE MY OWN MATERIAL?' p21.

Before any promotional material is given to you to use, it will have been certified by senior members of your company. To make sure that it complies with the Code, they will have checked:

Content
Provision of Prescribing Information
Use

WHAT CAN PROMOTIONAL MATERIAL CONTAIN?

Statements must be consistent with the SPC
This means in all respects, including indications, dosage, patient population, contra-indications, adverse events.

Statements must be accurate and balanced
It is possible for a statement to be accurate but not balanced – it could be 'the truth' but not 'the whole truth' e.g.

> A detail aid accurately presents the results of one study in which your product was superior to a competitor product. This would not be balanced if there was no mention of four similar studies which showed that the competitor product was superior to yours.

Statements must be up to date
This is particularly important in relation to costs and to SPCs, both of your company's products and of competitor products.

Statements must not be misleading

This is one of the most difficult areas in the Code. It depends very much on interpretation. Whether or not a piece is misleading is based on the overall impression it is likely to create.

A statement can be misleading because of:

- What is said – this may not be sufficiently clear, precise and unambiguous
- What is not said – there may be information which, if it was included, would alter the meaning or impression of the statement
- In what context it is said – the context can create a misleading impression

Statements must be capable of substantiation

Every statement made in promotion must be supported by data, which must be:

> Relevant
> Able to be provided i.e. not copyright or privileged
> Provided if requested without delay

Usually published references will be used, but unpublished data may also be used and referenced as 'Data on File'.

No hanging comparisons

A hanging comparison is a comparison which can be followed by the question 'compared to what?'

> e.g. 'Persil washes whiter'
> 'Nine out of ten cats prefer Whiskas'

In promotion it must always be clear what comparison is being made.

No superlatives without substantiation

Grammatically, a superlative is a term 'expressing the highest degree of a quality':

> e.g. 'the most effective'
> 'the best'
> 'the cheapest'
> 'the most widely prescribed'

Superlatives can be used only if they can be substantiated by data.

No exaggerated claims

i.e. a claim for a 'special quality' of a product which cannot be substantiated:

Unique
Unsurpassed
Outstanding
'the drug of choice'
'the gold standard'

These claims could be used only in the very few cases where they could be substantiated.

No product is 'safe'

It must not be stated or implied that any product has no side effects.

The word 'safe' (and related words e.g. safety) must not be used without qualification, 'well tolerated' and 'tolerability' are often used instead.

'New' for 12 months only

A product, presentation or indication can only be described as 'new' for 12 months from first marketing in the UK.

No competitor brand names

Brand names of other companies' products must not be used in promotional material without permission. (It is likely that use of brand names will be allowed when the Code is revised during 2001).

Brand names can be used in training and briefing material which is for your personal use only.

Comparisons must be fair

Comparisons with other companies' products must be fair, consistent with the SPC, accurate and not misleading. They must not be disparaging – i.e. no 'knocking' copy.

When comparing safety profiles, it must never be implied that another product is unsafe.

Cost comparisons must compare like with like

As with any claim in promotion, a cost comparison must be accurate, not misleading, up-to-date and consistent with the SPC.

Also, it must compare like with like, so that the comparison is fair:

> e.g. cost per day would not be a fair comparison of your product, prescribed as a seven day course, and a competitor prescribed as a five day course.

Comparators must be selected on a fair and valid basis:

> e.g. it is not fair to compare your company's product only with products which are more expensive

Any claim for cost effectiveness must take into account relative efficacy as well as all of the above.

Quotes are 'statements'

As far as the Code is concerned, quotes are judged in the same way as any other statements, so all of the above applies.

You cannot say something in a quote that you cannot say out of quotes. No matter how eminent the person quoted, if the quote itself is not accurate, balanced, fair etc it is in breach of the Code.

A quote must reflect the author's meaning and must represent the author's current view.

When a quote is not from a publication but from a personal communication, a medical meeting or similar, it can only be used with the permission of the speaker.

Graphs and figures must not be misleading

They must be accurate, clearly referenced, clearly presented and not visually misleading.

Promotion must not be likely to cause offence

Promotion must:

> **recognise the special nature of medicines**
> **recognise the professional standing of the audience**
> **be of a high standard at all times**

Artwork must not use naked or partially naked people or sexual imagery to attract attention to the material.

WHAT IS PRESCRIBING INFORMATION?

Prescribing Information (P.I.) is exactly that – information needed for prescribing

It must include:

- The name and list of ingredients of the product and its legal classification
- At least one indication
- Dosage and method of use
- Side effects, precautions and contra-indications
- Cost
- The name and address of the company which holds the Marketing Authorisation

P.I. and SPC are not the same thing

All the information in P.I. must be consistent with the SPC but the P.I. must include the cost, which is not in the SPC.

Promotional material must include Prescribing Information

Prescribing Information is needed on almost all promotional items which include:

- the brand name of a product

or

- the generic name and any other information about the product e.g. an indication, a claim

P.I. is not needed on:

- Promotional aids/giveaways which carry only the name of a product e.g. mugs, pens, clocks etc.

- material for your own personal use only e.g. training materials – one reason why you must not use this material in any promotional way (see p53).

P.I. must be part of the promotional piece not separate from it

This means e.g. on one of the pages of a detail aid, on the reverse of a letter.

The only exceptions to this, when P.I. may be given separately, are:

- Reprints
- Exhibition stands
- Audio-visual material
- Interactive data systems

If you are using these, you must follow the instructions you are given about how to provide P.I. Make sure that it is P.I. you provide, rather than the SPC – using a leavepiece which includes P.I. would be an acceptable way of doing this.

II THE MATERIAL

HOW CAN PROMOTIONAL MATERIAL BE USED?

You must use only current promotional material

Material may become out of date and hence no longer comply with the Code as a result of e.g. price changes, availability of new clinical data. Material which has been the subject of a complaint, may have to be withdrawn.

You must return or destroy material when instructed to do so. Do not hold on to 'favourite' pieces – continuing to use out of date or non-compliant material is a breach of the Code.

You should leave only leavepieces

Do not leave material which has not been designed to be left e.g. training material, detail aid intended for 'show only', photocopied material, as this may breach the Code.

You should give a promotional piece only to its intended audience

Promotional material should only be given to those who can reasonably be assumed to have a need for or interest in it.

Your company should instruct you about who should receive each piece e.g. GPs, GPs with an interest in cardiology, hospital cardiologists, practice managers.

It is important to follow instructions as material which may be acceptable for one group, may not be acceptable for another.

You should distribute a promotional piece as instructed

Promotional material can be distributed by mail or by hand e.g. during a call, at an exhibition stand, at a meeting but, as different pieces may need to be distributed in different ways, you should always follow any instructions your company gives you.

Promotional material cannot be distributed:

- **by e-mail**
- **by telex**
- **by fax**

unless prior permission has been given by the recipient.

You must not alter promotional material in any way

Even slight changes to a piece can mean that it no longer complies with the Code.

- **Do not write on the piece**
- **Do not use highlighter**
- **Do not use post-it notes**
- **Do not use only part of the piece**
- **Do not photocopy the piece**

CAN I PRODUCE MY OWN MATERIAL?

Many companies do not allow representatives to produce any material. Make sure you know your company's policy on representative-produced material.

Do not produce material	☐
Submit material to manager for approval	☐
Submit material to _____ for approval	☐

Any material you produce must comply with the Code

Any material you produce is almost certain to be considered promotional and must therefore comply fully with the Code in terms of:

Content – see p15
Provision of prescribing information – see p19

Representative produced material must be certified, to ensure that it complies with the Code. Because of the practical difficulties of this, many companies do not allow their representatives to produce any material at all. Others restrict the type of material which representatives are allowed to produce. Make sure you follow your company's policy on producing material.

Material produced for you by someone else must comply with the Code

Material produced by someone else on your behalf or at your suggestion is considered as though it is produced by you:

e.g. invitation letters from the chairman of a meeting
prescribing guidelines from a PCO to its members

If you have instigated, facilitated or distributed the material, it is considered promotional and must comply fully with the Code. Your company is responsible for ensuring that the content complies with the Code. The material needs to be approved.

Your company's involvement must be clearly stated:

e.g. 'Sponsored by...'
'on behalf of...'

II THE MATERIAL

CAN I WRITE A LETTER?

Many companies do not allow representatives to write any letters. Others may provide standard proforma letters for representatives to use. Make sure you know your company's policy on representative letters.

- Do not write any letters ☐
- Use standard letters only ☐
- Submit letters to manager for approval ☐
- Submit material to _____ for approval ☐

Any letter you write must comply with the Code

Any letter you write is likely to be considered promotional and must therefore comply fully with the Code.

This means in terms of:

Content – see p15
Provision of P.I. – see p19

P.I. is required on any letter which contains the brand name of a product

or

the generic name and any other information about the product e.g. an indication, a claim

P.I. must be on the reverse of the letter, not separate from it

An answer to an individual enquiry is not considered promotional

A letter written in response to an individual enquiry is not considered promotional provided that the response is:

- Specific to the query
- Accurate
- Not misleading
- Not promotional in nature i.e. in content, in presentation or in use

It is likely that the letter will be considered promotional simply because it is written by a representative. It is therefore preferable for a letter like this to be sent by Medical Information.

A letter produced for you by someone else must comply with the Code

A letter produced by someone else on your behalf or at your suggestion is considered as though it is produced by you:

e.g. invitation letter from the chairman of a meeting

If you have instigated, facilitated or distributed the letter, it is considered promotional and must comply fully with the Code. It needs approval and certification.

CAN I SPONSOR MATERIAL?

In many companies, all sponsorship is handled by Head Office. You should be aware of your company's policy on sponsorship:

- Submit all requests to Manager ☐
- Submit all requests to _____ ☐

Any material you sponsor will be subject to the Code, regardless of whether it is promotional or educational

Note that all of the following would be considered 'sponsorship':

- Provision of grant to produce a patient booklet
- Payment for printing of a practice leaflet
- Copying in house of a Chairman's letter to PCO members

You can only sponsor material if it could be considered to enhance patient care or benefit the NHS.

If you have any input at all into the content of the material, your company will be responsible for this

Your company is responsible for the content of the piece –

UNLESS

you have no input, control or knowledge of the content prior to publication.

If the material is promotional, either in terms of content or how it will be used, it must comply fully with the Code, including the need for P.I.

It must be made clear that the material has been sponsored by your company

This applies whether the material is promotional or educational.

Failure to do this can result in it being considered disguised promotion.

CAN I USE REPRINTS?

You should use reprints only as instructed

Your company must decide whether or not a reprint can be used.

For a reprint to be acceptable:

> the paper must have been refereed
> the content must comply fully with the Code

This applies to reprints you give during a call and also to reprints you make available on an exhibition stand.

P.I. must be provided with a reprint

It can be provided as a separate document – the only form of printed material where the P.I. does not have to be part of the piece itself.

You can provide a health professional with a reprint he has requested

A reprint which is requested is outwith the scope of the Code, as long as you do not use it in a promotional way.

BUT

You must not photocopy a reprint

The number of copies of a paper which a company can make is controlled, not by the Code, but by the Copyright Licensing Authority – there are substantial financial penalties for companies who do not comply with these. Follow instructions about whether or not you are able to photocopy material. If not, you should ask Medical Information to provide another copy.

You must not photocopy any paper which has been given to you for personal training purposes only.

CHAPTER III – THE CALL

WHO CAN I CALL ON?

You can promote to health professionals

'Health professional' is defined as anyone who may prescribe, supply or administer a medicine, so this includes:

- Doctors
- Dentists
- Pharmacists
- Nurses

You can promote to appropriate administrative staff

'Administrative staff' includes:

- Practice Managers
- Health Authority Managers
- PCO members

'Appropriate' means that you must be able to justify the promotion as relevant to that person's work:

> e.g. promotion to Practice Managers could be justified if it was relevant to the running of the practice
>
> promotion to Health Authority managers or to PCO members could be justified if there were budgetary or management implications

You cannot promote to the General Public

The Code reflects UK law which specifically prohibits promotion to the general public.

If you encounter members of the public in administrative roles within the NHS, you may promote to them as 'administrative staff' but only if you can justify the promotion as relevant to them in that role.

Patients are considered as members of the public. If you have any contact with patients, you must make sure that nothing you do or say could encourage them to seek a prescription for any particular medicine. You must never discuss individual medical matters with a patient.

WHEN CAN I CALL?

Calls must not be inconvenient

You need to consider:

- Frequency (see below)
- Duration
- Interval between calls
- Timing

You should respect the wishes of any individual and you must follow any local requirements or procedures e.g. in hospitals.

Always treat any individual's time with respect. Make sure you are punctual and that you give adequate advance notice if you have to cancel or change an appointment.

HOW OFTEN CAN I CALL?

Three calls per year per doctor

The Code states that 'the number of calls made on a doctor by a representative each year must not normally exceed three on average'.

The following contacts do not count towards the total:

- Attendance at a group meeting
- Any visit made at the doctor's request
- Any visit made to respond to a specific enquiry
- Any visit made to follow up an adverse event report

You could thus see a doctor more often than three times a year and still be within the Code.

HOW CAN I ARRANGE A CALL?

By letter
Any letter you write, even a purely administrative one, will fall within the scope of the Code. If your letter mentions a product it is likely to be considered promotional and must comply with all Code requirements. Make sure you are aware of your company's policy on letters. See p22 for details.

By phone, e-mail, fax
The Code prohibits promotion via phone, e-mail or fax, unless the recipient has given permission in advance. You can use these methods for administrative purposes only e.g. making or confirming appointments, but make sure that your call or message is not in any way promotional.

Identify yourself and your company clearly
Always be clear about who you are and which company you represent. Never use any subterfuge to gain an interview e.g. do not describe yourself as a 'personal caller'.

Maintain a high standard of ethical behaviour at all times.

Do not offer or pay a fee for an interview
You must not pay a fee either to an individual health professional or to a practice or departmental fund.

It is the guidance of the GMC to doctors that they must not ask for or accept a fee in return for granting an interview as this is against the law. If you are asked for a fee, you should report this to your manager. Your company can ask the PMCPA to take the matter up with the doctor (without naming you or your company).

Do not use any inducement to gain an interview

The following would be considered inducements:

> Making a donation to charity in return for interview
>
> Providing a book for the library in return for interview
>
> Insisting on delivering a requested item to a doctor, rather than leaving it with the receptionist

Delivering an item:

> **YOU:** *"Can I see Dr Smith?'*
> **Receptionist:** *"He's very busy this afternoon"*
> **YOU:** *"I have the Dermatology Atlas which he requested on this card"*
> **Receptionist:** *"He could see you next Thursday – 3 o'clock?"*
> **YOU:** *"OK. I'll bring the Atlas back then"*

You have just broken the Code.

> Delivering an item offers you an opportunity to gain access to a doctor but if you insist on delivering it only to the doctor, you are using it as an inducement to gain an interview
>
> It has been judged in previous Code of Practice cases that if, as in this case, you make it known that you have an item to deliver, you must leave it, whether or not you are specifically asked to do so

III

THE CALL

CHAPTER IV – THE DETAIL

HOW CAN I DETAIL?

You can only detail face to face

You cannot detail via phone, fax or e-mail, without permission in advance from the person you wish to detail.

WHAT CAN I DETAIL?

You can only promote a product or indication which has a UK product licence

If you are asked about an unlicensed product or indication, you should explain that it is not possible for you to discuss it. You could offer to refer the question to Medical Information, who should be able to deal with it (see p54).

There are two specific situations in which you may be able to discuss a product or an indication which does not have a product licence.

International conferences

At an international conference held in the UK, it is possible to promote a product or indication which is not licensed in the UK, but is licensed in another country, subject to various specific requirements.

Budget holders

If a product or indication is likely to have significant budgetary implications, information about it can be given, before it is licensed, to Health Authorities, Trust Hospitals and Primary Care Organisations. This is to facilitate financial planning. In such cases:

Information can be sent only to those responsible for making policy decisions on budgets – not to those expected to prescribe

Information must be factual and non-promotional

Your company will decide if either of these circumstances applies and whether or not it is appropriate for you to be involved. If so, it is important that you follow briefing instructions exactly.

WHAT CAN I SAY?

The Code requirements for the spoken word are essentially the same as for the written word. These are set out in detail on page 15.

You should always follow any script or briefing you are given by your company.

Whether or not you have a script or briefing, you should bear the following in mind.

What you say must be consistent with the SPC

If, during the course of a detail, you are asked something about the product which is not consistent with the SPC e.g. an unlicensed indication, a route of administration which is not in the SPC, you should explain that it is not possible for you to discuss it. You could offer to refer the question to Medical Information, who will be able to deal with it (see p54).

What you say must be accurate and balanced

Make sure that if, perhaps through pressure of time, you are not able to complete a detail, you do not omit information which is important for providing balance.

What you say must not be misleading

Again make sure that pressure of time, or less than ideal conditions, do not result in what you say being misleading. This could happen if:

What you say is not sufficiently clear, precise and unambiguous

You do not give all the information which should be given e.g. you present only the efficacy results and miss out the safety analysis

You say something out of context

What you say should not include superlatives or exaggerated claims

Although these are frequently used in normal conversation, you must be careful not to use them in a detail.

Also make sure you don't use 'safe' or 'safety' – 'well-tolerated' and 'good tolerability' may be more of a mouthful but will keep you within the Code.

IV THE DETAIL

What you say must not disparage

**another company
the product of another company
any health professional**

Beware the throwaway comment!

Never say anything which suggests that another product is unsafe or ineffective. You should not use the brand names of other companies' products. (This is likely to be allowed after the Code is revised during 2001).

What you say must be capable of substantiation

It may be tempting to add information into a detail, but there must always be evidence to support it:

**e.g. 'This is the most commonly prescribed NSAID at the General Hospital'
Do you have prescribing data to support that statement?
Do you have permission to use it?
Would you be able to provide it in writing, if asked?**

This is particularly important if you are quoting someone, e.g. *'Dr Smith at the General says this NSAID is the one to choose for elderly patients'*:

**Are you quite sure that he does?
Has he given you permission to quote him?**

WHAT DO I NEED?

You must have available a copy of the SPC for any product you detail

'Available' means in your bag – not in the car or at home – so that you can give a copy to anyone who asks for it.

You don't need to have an SPC for a product which you are asked about, but you will probably want to arrange to provide one later, either yourself or via Medical Information.

You don't need to have an SPC for a competitor product of another company.

CHAPTER V – THE GIFT

WHAT GIVEAWAYS CAN I USE?

'Giveaways' i.e. mugs, pens, clocks etc. Also called 'promotional aids', 'promotional gifts', 'access items'.

Giveaways must be inexpensive

'inexpensive' currently (January 2000) means no more than £5.00, excluding VAT, cost to the company.

Giveaways must be relevant to the practice of medicine

This can be difficult to interpret.

Items clearly for medical use are acceptable e.g. tourniquet, blood pressure cuff, surgical gloves.

Items clearly for home use are not acceptable e.g. table mats, plant seeds.

Items which could have both medical and home use are more difficult e.g. clock, phone card, toys for children.

Just because a doctor may use an item, does not mean that the item is 'relevant to medicine' e.g. road atlas, eyeshades are not acceptable.

Giveaways for patients are acceptable if they are inexpensive and related to the patient's condition or to general health.

Use only giveaways supplied to you by your Company

These will have been approved as complying with the Code.

Do not buy gifts. Boxes of chocolates and bottles of wine are not considered relevant to the practice of medicine – even at Christmas!

Distribute giveaways only as instructed by your Company

If a giveaway is intended for hospital consultants, do not give it to nurses.

A giveaway may be acceptable for one group of health professionals but not for another.

There must be no conditions attached to a giveaway

The giveaway must not be, or seem to be, an inducement to prescribe i.e. the offer of the giveaway must not depend on the doctor using a particular product. Even though giveaways are sometimes referred to as 'access items', they must not be used as an inducement for an interview.

Insisting on seeing a doctor to deliver an item would mean that you are using it as an inducement to gain an interview – if a doctor is not willing to see you, you must leave the item (see page 28). It is for this reason that all offers of giveaways will state, usually on the RPC *'This offer carries no obligation to see a representative'* or similar.

CAN I BE INVOLVED IN A COMPETITION?

Your company is responsible for all competitions and prizes with which you are involved

This is true whether you are involved directly or through a third party:

> e.g. If you donate a prize to a competition organised by a Hospital Department, your Company will be responsible if either the competition or the prize does not comply with the Code

Competitions must be professional

They must be of a high standard, a test of skill and relevant to medicine or pharmacy. They must also be in good taste, and not likely to cause offence.

Competition prizes must be inexpensive

A prize can be up to a maximum of £100.00, excluding VAT, if the competition is a serious one and if the prize is not out of proportion to the skill required.

The number of prizes should be appropriate for the number of possible entries, the actual number of responses and the difficulty of the competition.

CAN I PROVIDE MEDICAL AND EDUCATIONAL GOODS & SERVICES?

'Medical and educational goods or services' are goods or services which can be considered to enhance patient care or benefit the NHS e.g.

- **medical equipment**
- **medical textbooks**
- **audit facilities**
- **training courses**
- **financial support for clinics and/or clinic staff**
- **patient education programmes**

In many cases it will not be appropriate for you, as a sales representative, to have any involvement with medical and educational goods or services

Any medical and educational goods or services which a company provides must be non-promotional

'non-promotional' means:

- they must not bear the name of a product – they may bear a company name
- they must not be linked to the use of a product
- they must not be or appear to be an inducement to prescribe
- any associated materials must be non-promotional

They must not provide any personal benefit to the recipients.

Any involvement you have must be wholly non-promotional

If your company decides that it is appropriate for you to be involved, you will be given detailed instructions which you must follow exactly.

If you do not, there is a risk that the goods or services will be considered promotional and in breach of the Code.

When providing, delivering or demonstrating goods or services, you must not link this in any way with promotion.

When delivering a medical or educational item, you must not use it as an inducement to gain an interview – see p28.

You must have no direct contact with patients or with patient data

Patient confidentiality is of critical importance therefore:

> **You should not sit in on clinics, or be present during screenings.**
>
> **You should not be involved in any discussion on individual clinical findings.**
>
> **You must have no access to any data or records which could identify, or could be linked to, particular patients e.g. you should not be involved in an audit of patient records.**

V

THE GIFT

CAN I SPONSOR THIS?

In many companies, all sponsorship is handled by Head Office. You should be aware of your company's policy on sponsorship:

Submit all requests to Manager ☐
Submit all requests to _____ ☐

If your company policy allows you to be involved in providing sponsorship, be aware of the following:

Anything you sponsor is subject to the Code

'Sponsorship' includes:

payment, in full or in part
provision of a grant
provision of a service e.g. printing a booklet, delivering meeting invitations

Under the Code, sponsorship must enhance patient care or benefit the NHS

It must always be made clear that the material or activity has been sponsored by your company

This applies whether the material or activity is promotional or educational. Failure to do this can result in the material or activity being considered disguised promotion.

Sponsorship must never be an inducement to prescribe

Sponsorship must not be linked in any way to the prescription of a product.

Sponsorship should be provided in the name of your company and not linked to any product.

CAN I SPONSOR MATERIAL? See p23.

CAN I SPONSOR A MEETING? See p51.

V THE GIFT

CAN I MAKE A DONATION?

In many companies, all donations are handled by Head Office. You should be aware of your company's policy on donations:

Submit all requests to Manager ☐
Submit all requests to _____ ☐

If your company policy allows you to be involved in donations, be aware of the following:

Any donation you make in the course of your work will be considered subject to the Code

Effectively, this means any donation which goes through your expenses. The Code does not cover donations you make as a private individual.

A donation must be to a registered, reputable charity

Any contributions to other funds e.g. a departmental travel fund or to a practice equipment fund should be considered as a Medical or Educational service – see p35.

A donation must not be an inducement to prescribe

The donation must not be linked in any way to the prescription of a product.

The donation should be made in the name of your company and not linked to any product.

A donation must not be an inducement to grant an interview

You cannot make a donation in return for an interview.

Nor can you make a donation in return for being given the opportunity to host or present at a meeting.

At meetings, it may be possible to make a small donation in return for a health professional visiting your company's stand, but this should be arranged by your company.

CAN I OFFER A PACKAGE DEAL?

Package deals must be fair and reasonable

Any benefit offered as part of a package deal must be relevant to the medicine purchased

- e.g. providing the equipment necessary to administer a product
  providing training on the disease area in which the product is used

CAN I OFFER A DISCOUNT?

Discounts are not usually covered by the Code but other trade practices may be.

The benefits of discount schemes must be acceptable

They must not bring any personal benefit to health professionals e.g. gift vouchers for High Street stores are not acceptable, even if they are an alternative to financial discounts.

Any item given as part of a discount scheme must be relevant to the practice of medicine – see p33.

V THE GIFT

CAN I GIVE SAMPLES?

Samples are clearly defined in the Code
A sample is a small supply of a medicine provided for identification purposes or to allow health professionals to familiarise themselves with its use.

- It must be no larger than the smallest presentation on the market.
- It must be marked 'Free medical sample – not for resale' or similar.

Anything else is not a sample.

You should only use samples which have been provided to you by your company for this purpose.

You can give samples only to health professionals who are able to prescribe the product
You cannot give samples to administrative staff or to members of the general public.

You can give samples only in response to a written request
The request must be signed and dated by the health professional – a pre-printed form can be used.

The number of samples you can give is limited
You can give a health professional no more than 10 samples of any one medicine per year.

You must deliver the sample correctly
You must hand the sample directly to the health professional who requested it, or to someone he has authorised to receive it on his behalf.

You should not use samples as an inducement to gain an interview. (See p28).

If you distribute samples in hospitals, you must comply with that hospital's requirements.

You must give a copy of the SPC with the sample.

You must store samples securely
You must take all reasonable steps to prevent inappropriate or inadvertent use of samples. Do not store samples in your car.

Samples must be recorded
Your company must have a system of control to ensure that all the above requirements are met and must retain all sample requests for at least one year.

CHAPTER VI – THE MEETING

You should always follow your company's policy on meetings.

WHAT SORT OF MEETING CAN I HOLD?

Any meeting you hold must have a clear educational content. 'Educational content' is judged in terms of:

- quality
- quantity
- relevance to the attendees

PGEA approval is good evidence of educational content but will not guarantee that a meeting is wholly acceptable under the Code.

Education must be the main purpose of the meeting

Hospitality therefore must be secondary.

To get the right balance between education and hospitality, you need to consider the following:

Time
- how many hours of education and how many hours of hospitality.

Attractiveness of hospitality
- in relation to that of educational component.

Attractiveness of venue
- in relation to that of educational component.

A very exclusive or luxurious venue could appear to be the main attraction of the event.

Arrangements
e.g. 'discussion over dinner' suggests that dinner is the more important part of the evening – you should make sure that education is clearly separate from the hospitality.

Emphasis
Material associated with the meeting e.g. invitation, programme, poster etc should not seem to 'sell' the meeting on the venue or hospitality, rather than on the educational value. Don't use phrases like these:

> *'Gala Dinner'*
> *'Champagne reception'*
> *'elegant hotel situated in beautiful grounds'*
> *'a great night out'*

CPD/PGEA approval
This is good evidence of educational content, but does not mean that the meeting is wholly acceptable under the Code. It is the overall balance between education and hospitality which is important.

WHO CAN ATTEND A MEETING?

Meetings should only be attended by health professionals and appropriate administrative staff

Whether or not it is appropriate for administrative staff to attend should be determined by whether or not the educational content of the meeting is relevant to them:

> e.g. a practice manager could be invited to 'Audit in General Practice' but probably not to 'Recent Advances in Paediatric Surgery'

Spouses or other accompanying persons must not be invited to attend a meeting –

> unless that person is a health professional or appropriate member of administrative staff who qualifies to attend the meeting in his/her own right. If that is the case, the accompanying person should attend both the educational and social parts of the meeting

If any accompanying person attends a meeting uninvited, all costs must be paid for by the person he/she is accompanying.

WHERE CAN I HOLD A MEETING?

The venue must be professional and appropriate to the occasion

This obviously rules out nightclubs, casinos etc. You should also be careful if you are using conference facilities which are part of sporting facilities e.g. golf clubs, football clubs, amusement parks. Make sure that all the other requirements in this section are met – in particular, make sure that the venue does not appear to be the main attraction of the meeting. See also p41.

The venue should offer privacy for the educational part of the meeting

If the meeting is in a restaurant, the educational part should be held in a private room or area.

You must be able to justify a distant venue

You could justify holding a meeting for GPs from all parts of UK almost anywhere in UK as it is obvious that significant travel will be necessary for many attendees wherever it is held. It would be difficult to justify holding a meeting for GPs from Devon anywhere other than the southwest of the country.

The cost of the venue must not be excessive

The cost should be included in the overall cost of hospitality, which should not exceed what the attendees would pay for themselves.

It is the overall impression which counts

A venue which appears exclusive or luxurious is not acceptable, even if you get a good deal so that the actual cost is not excessive.

The venue must not seem to be the main attraction of the meeting. See p41.

Your company may consider some specific venues or types of venue unacceptable. If so, you can list them here.

VI

THE MEETING

WHAT HOSPITALITY CAN I OFFER?

The following applies to hospitality provided to Health Professionals or Administrative Staff. It does not apply to other categories e.g. wholesalers.

Hospitality must be appropriate to the occasion

A simple sandwich lunch would be appropriate if you are giving a product presentation to a GP practice or hospital department.

A formal dinner could be appropriate after a state of the art lecture by an eminent authority to a group of consultants.

Hospitality must be appropriate to the attendees –

i.e. no more than the attendees would be expected to pay for themselves.

Many companies have limits on the amounts which can be spent on lunch and dinner. You should always keep within your own company's limits but be aware that on many occasions, the maximum amounts will be inappropriate.

Maximum allowable cost per head:

Lunch _____

Dinner _____

Hospitality must be appropriate to the arrangements for the meeting

Meals and accommodation must be justified by the programme of the meeting and by practical necessity:

> e.g. For a meeting which starts first thing on Saturday morning, it would be reasonable to offer dinner and accommodation on Friday night if the attendees need to travel some distance, but it would not be reasonable to offer this for a local meeting where all attendees live near the venue

The impression created is all important

Hospitality which appears inappropriate or excessive is unacceptable, even if the actual cost e.g. as the result of a special deal is within reasonable limits.

WHAT ACTIVITIES CAN I OFFER?

Any activity you offer must be professional
In the past, golf and wine tastings have been ruled to be unprofessional.

You cannot offer any sporting activity as part of a meeting
This means sporting activity either as a participant or as a spectator

Sporting activity should not form part of the official programme of the meeting.

Sporting activity should not seem to be the main attraction or purpose of the meeting.

It may be acceptable for meeting attendees to take part in sporting activity after the meeting has officially ended or during breaks in the meeting:

- e.g. after a Friday evening and Saturday morning meeting attendees may play golf on the hotel course but:
    - this should not be included in the programme of the meeting
    - this should not be advertised as a feature of the meeting
    - this should not be organised by you or your company
    - they must pay for all costs themselves
    - if you play golf with the attendees, you should pay your own costs, not claim them through expenses

> REMEMBER THAT IT IS THE IMPRESSION WHICH IS IMPORTANT.

VI THE MEETING

WHAT MEETING MATERIALS CAN I USE?

All materials associated with the organisation or content of a meeting must comply with the Code

This includes:

- Invitation
- Programme
- Welcome letter
- Poster
- Presentations
- Briefing for speakers
- Lecture notes for attendees
- Meeting report

P.I. may be required on meeting materials

P.I. is required on any meeting material which includes:

- the brand name of a product

or

- the generic name and any information about the product

Meeting materials produced by third parties must comply with the Code

e.g. invitation from chairman
welcome letter from meeting organiser
handouts from external speaker
meeting report by faculty

All meeting materials must clearly state that your company has sponsored the meeting

This applies whether the meeting is promotional or educational.

The impression is important

Meeting materials should not give the impression that the meeting does not comply with the Code, in particular regarding:

- **Main purpose of the meeting**
- **Educational content**
- **Cost of hospitality**
- **'Lavishness' of hospitality**
- **Availability of sports facilities**

The impression should always be professional.

CAN I HAVE A STAND AT A MEETING?

You can only have a stand at a meeting which complies with the Code

The meeting must be primarily educational and all hospitality must be acceptable. This applies even if you are not directly responsible for the meeting.

You should obtain permission for a stand

Some venues e.g. Postgraduate Centres, have rules about promotional stands at meetings so always check with the venue and/or the organisers of the meeting whether – and where – a promotional stand is acceptable.

You should use only approved promotional material on the stand

Reprints available on a stand must comply with the Code – see p24.

A stand promoting products should not be accessible to the public

If the venue is a public place such as a hotel, you should position the stand so that it is not easily accessible to the general public i.e. not in the main lobby.

VI

THE MEETING

CAN I MAKE A PRESENTATION AT A MEETING?

Any presentation you make must comply fully with the Code
Details of the requirements can be found on p15 and 30.

You should therefore only use material which has been provided to you by your company. This will have been approved and certified to ensure that it complies with the Code.

You should use the presentation only as instructed e.g. don't use only part of the presentation if you are instructed to use it as a whole.

You should make the presentation only to the categories of people specified e.g. if you are told to present it only to hospital specialists, do not show it to general practitioners.

You must follow instructions about providing Prescribing Information
Usually, this will mean having copies of P.I. available at the meeting for anyone who asks for it. You can use any leavepiece which includes P.I. This also applies if you are showing a video or film.

You must make it clear that the presentation is a promotional one
In many situations this will be obvious, but make sure that the presentation is not described in invitations etc in any way which might suggest that it is not promotional. Make sure that a promotional presentation is acceptable to the organisers of the meeting – this may not be the case at some PGEA meetings.

CAN I INVITE A SPEAKER TO MAKE A PRESENTATION?

Your company is responsible for ensuring that all content of a meeting, including a presentation made by an invited speaker, complies with the Code

This means that the presentation should be:

Accurate and objective
Consistent with SPC
Balanced
Capable of substantiation

It is usually not possible to have a speaker's presentation approved before a meeting, and, in any case, it is not appropriate for a company to control the content of a speaker's presentation. Speakers are entitled to express their own opinions.

Nevertheless, it is expected that a company will be familiar with a speaker's views and opinions and will therefore know whether or not the speaker is likely to stay within the Code.

You should be aware if it is likely that the speaker's presentation will not comply with the Code

When the speaker is from your area, you will be in the best position to know his views and opinions on the subject of his presentation. You should not use a speaker if you feel that it is likely that he will not stay within the Code.

You cannot select a speaker because he is a supporter of your product, unless his support can be substantiated by data

Any briefing or information you give the speaker must comply with the Code

If you provide the speaker with any information to use in his presentation, you should make sure that it complies with the Code as your company will be considered responsible for this.

VI
THE MEETING

WHAT CAN I PAY FOR AT A MEETING?

All payments made in relation to a meeting are subject to the Code

Venue
You can pay room rental to postgraduate medical centres and the like but cannot make any payment to doctors or groups of doctors for rental of rooms.

Travel
You can pay for travel for attendees if the meeting takes place outside their normal working area. You should not pay for travel to local meetings.

You can either reimburse the costs of travel against receipts or you can provide travel in the form of train tickets, coach service etc.

Hotel accommodation
You can pay for accommodation only if this is justified by programme.

> e.g. If a meeting starts on a Friday and finishes at lunchtime on Saturday, you could justify accommodation for Friday night, but not for Saturday night

You cannot justify, and therefore cannot pay for, accommodation for a local meeting.

Meals
You can pay for meals which are justified by the programme of the meeting.

The cost should be as discussed above under hospitality i.e. no more than the attendees would pay for themselves.

Many companies have limits on the amounts which can be spent on lunch and dinner. You should always keep within your company's limits but be aware that on many occasions, the maximum amounts will be inappropriate.

Maximum allowable cost per head:

> Lunch _____
>
> Dinner_____

To calculate the cost per head, you should include all costs of the hospitality, including room hire if relevant, and divide this by the number of **intended** attendees i.e. if you arrange catering for 60 but only 30 turn up, you divide the total cost by 60. Make sure you document the costs in detail.

You cannot make any payment instead of providing a meal e.g. you cannot make a donation to the Practice Fund instead of providing lunch.

If food is provided at a meeting, you should obtain an invoice for any payment you make. The amount of the invoice must be reasonable given the food provided e.g. an invoice for

£100 would not be **reasonable** for a lunch of supermarket sandwiches and soft drinks for 10 people – even though the amount per head (£10.00) is **acceptable.**

Activities
You can only pay for activities which are professional and appropriate to the occasion.

You cannot pay for sporting activities.

Speakers
You can pay an honorarium which is appropriate to the occasion and in keeping with usual professional rates. You should not pay an honorarium to a speaker within his usual working area e.g. a consultant presenting to his department, a GP presenting to his practice staff.

You can pay travel, accommodation and meals as above.

You can pay reasonable out of pocket expenses e.g. preparation of slides, provision of a locum.

Accompanying persons
You cannot pay any costs of accompanying persons or provide them with any hospitality. All costs must be paid by the persons they are accompanying.

Donations
You cannot make a donation in return for being given the opportunity to host or present at a meeting as this would be considered an inducement to grant an interview.

CAN I SPONSOR A MEETING?

You should always follow any policy your company has about sponsoring meetings.

If you sponsor a meeting, it is subject to the Code
'Sponsoring' includes any of the following:

- Paying all or some of the costs of a meeting
- Paying for a stand at a meeting
- Providing a grant for a meeting

This applies whether the meeting is promotional or educational.

You can only sponsor a meeting which is relevant to the practice of medicine
The meeting must be able to be considered as in some way enhancing patient care or benefiting the NHS.

If you sponsor a meeting, you are responsible for the meeting's content

You must ensure that the content complies with the Code.

The only exception to this would be if you sponsored a meeting in which you had no involvement or control e.g. a meeting of an independent working party.

Sponsorship must be clearly declared at the meeting itself and on any material relating to the meeting

e.g. posters, invitations, welcome letters etc

This applies whether the meeting is promotional or educational.

If you sponsor a meeting, the hospitality must comply with the Code

This applies whether the meeting is promotional or educational. See Hospitality p44.

INTERNATIONAL MEETINGS

At international conferences held in the UK, all promotional materials and activities, including hospitality, must comply fully with the Code

The only difference is that it may be possible to promote a product or indication which does not have a UK product licence if the product or indication is licensed in another country (subject to various requirements).

Hospitality abroad is subject to the same requirements as hospitality in the UK

At international conferences held outside the UK, all promotional material and the meeting content should comply with local requirements but all hospitality for UK health professionals must comply with the UK Code. This applies whether the hospitality is organised by the UK company or internationally.

CHAPTER VII – THE ESSENTIALS

TRAINING

Your company must give you training on the ABPI Code of Practice and on the products you promote

Your company should provide training on the technical aspects of each medicine you promote and should also give you briefing instructions on how it should be promoted.

Training and briefing material must not be used promotionally

All training and briefing material is subject to the Code and, like promotional material, is approved by your company but it can differ from other promotional material:

- It may include information on unlicensed products or indications, provided for your use only, as background information
- It does not need to include P.I.
- It may include brand names of other companies' products

Because of these differences, all your training and briefing material should be clearly labelled 'For Representative Use Only', 'For Personal Use Only' or similar.

- Do not give the material to any third party
- Do not show the material to any third party
- Do not copy the material

ABPI REPRESENTATIVES' EXAMINATION

You must pass the ABPI Representatives' Examination

The Examination covers basic subjects relevant to your work e.g. anatomy and physiology, disease, pharmacology and also information on the National Health Service and the pharmaceutical industry.

You must pass the Examination within two years of beginning to work as a representative – this means two years in total, not two years with any one company. You should be entered for the examination within your first year. Your company should provide you with appropriate training to pass this exam.

MEDICAL INFORMATION

Every company should have a scientific service responsible for information received about the company's products. This service will usually also be responsible for providing information about the products i.e. a Medical Information Service.

Contact details for Medical Information:

Phone:

Fax:

A letter from Medical Information is not covered by the Code, as long as:

the letter is in response to an enquiry from a health professional
the reply is accurate, not misleading and specific to the enquiry
the letter is not promotional in content, style or use

Such a letter may include information which could not be included in promotional material e.g. information about unlicensed products or indications and does not need Prescribing Information.

If any of these conditions is not met, the letter is covered by the Code and must meet all requirements, including the need for P.I. As a representative, you should bear in mind that:

a letter sent out under your name is likely to be considered promotional
detailing from a letter is likely to be considered promotional use
you must not use a letter as an inducement to gain an interview

ADVERSE EVENTS

Reporting serious* adverse reactions is a legal obligation

Your company has a legal obligation to report all serious adverse reactions which occur with any of its products to the licensing authority. Reporting must be done within 15 days of receipt by a member of the company.

You must be aware of the procedure for reporting adverse events in your company so that, if a health professional mentions a serious adverse event to you, you are able to pass on the information quickly to the appropriate person.

Contact details for reporting adverse events:

Name:

Phone:

Fax:

This must be done IMMEDIATELY to allow time for full details to be obtained from the doctor and for the event to be reported to the licensing authority within the required time frame.

All adverse events are important

It is also important to pass on information about adverse events which are not classified as serious, particularly when these occur with new products, which carry a black triangle. Again you should be aware of your Company's procedure for doing this.

*serious – fatal, life-threatening, disabling, incapacitating or which results in hospital admission, prolongs hospitalisation for in-patients, is a congenital abnormality or is an important medical event

EXPENSES

You should make sure that you always follow your company's procedure for recording expenses.

If there is a Code complaint about any of your activities as a representative, your expense forms may have to be submitted to the PMCPA. Any item relating to hospitality, payments to doctors for any reason will definitely be within the scope of the Code and other items may also be relevant.

Contact details for expenses:

Name:

Phone:

Fax:

GLOSSARY

ADMINISTRATIVE STAFF – healthcare staff who do not qualify as Health Professionals but whose position means that they have a valid interest in or need of information about medicines e.g. Health Authority Managers, Practice Managers, PCO members.

ADVERSE EVENT – an unexpected or unpredicted reaction to a drug, unrelated to the drug's usual effect.

ABPI – The Association of the British Pharmaceutical Industry, the trade association representing manufacturers of prescription medicines. It represents over 70 companies which produce 80% of the medicines supplied to the National Health Service.

BLACK TRIANGLE – an inverted black triangle must be shown on promotional material for new medicines, to indicate that special reporting of adverse reactions is required.

BRAND NAME – the unique name used in marketing a product.

BMA – British Medical Association.

CERTIFICATION – the final approval of promotional material by senior members of a company, who certify that the material complies with the Code.

CODE OF PRACTICE PANEL – the body within the PMCPA which first considers and rules on complaints. The members of the Panel are staff of the PMCPA.

CODE OF PRACTICE APPEAL BOARD – the Appeal Board considers cases where the ruling of the Code of Practice Panel has not been accepted, either by the complainant or the respondent. The Appeal Board consists of industry and non-industry members.

CPD – Continuing Professional Development (see PGEA).

CONTRA-INDICATION – diseases, conditions or situations in which, for safety reasons, a drug must never be used.

GMC – General Medical Council, legally responsible for registering doctors and maintaining professional standards. The GMC considers complaints against doctors and decides whether or not the doctor should be 'struck off' the register.

GENERIC NAME – the official name(s) of the active ingredient(s) of a medicine. If a medicine is marketed by more than one company, it may have more than one brand name but the generic name will always be the same.

HEALTH PROFESSIONAL – professionals who may prescribe, supply or administer medicines e.g. doctors, dentists, pharmacists, nurses.

INDICATION – a disease or condition for which a drug is approved for use.

LEGAL CLASSIFICATION – medicines are legally classified as 'POM' (available on prescription only), 'P' (can be sold in pharmacies only) or 'GSL' (General Sales List – can be sold in any shop). 'P' and 'GSL' medicines are usually referred to as 'OTC' (over the counter).

MARKETING AUTHORISATION – must be granted by the MCA before a company is allowed to market a medicine.

MCA – Medicines Control Agency, part of the Department of Health, responsible for authorising the marketing of medicines and for monitoring all aspects of their use.

OTC MEDICINE – a medicine which can be advertised to the public and 'over the counter' (hence OTC) without a prescription. OTC medicines can also be prescribed by doctors.

PGEA (postgraduate education allowance) – an allowance paid to GPs if they undertake continuing medical education of 5 days per year on average. This is being replaced by CPD (Continuing Professional Development).

PGEA APPROVAL – granted by Postgraduate Deans if a meeting, presentation or course meets the required standards for continuing medical education.

PRESCRIBING INFORMATION (P.I.) – the essential facts about a medicine, based on the SPC. The contents of P.I. are specified by the Code (see p19).

PMPCA – Prescription Medicines Code of Practice Authority, the body set up by the ABPI to administer the Code.

PCO – Primary Care Organisation, this term includes PCGs (Primary Care Groups), PCTs (Primary Care Trusts) etc. Multidisciplinary groups responsible locally for development of primary and community healthcare. Groups may have budget responsibility. Members include GPs, nurses, Health Authority personnel and members of the public.

PRODUCT LICENCE – see Marketing Authorisation.

PROMOTIONAL AID – an inexpensive item carrying only the name of a medicine and/or a company e.g. mug, pen, clock (see p33).

REFEREED – Most scientific journals publish papers only after they have been refereed, i.e. have been reviewed by experts in the field.

REPRINT – a printed copy of a published paper, usually ordered and purchased from the journal in which the paper was published.

RPSGB – Royal Pharmaceutical Society of Great Britain.

SAMPLE – a small supply of a medicine provided so that Health Professionals can familiarise themselves with it. (see p40).

SPC or SmPC – Summary of Product Characteristics. Detailed information about all aspects of a medicine relevant to its use, including dose, indications, contra-indications, adverse events. The SPC is agreed with the MCA prior to the granting of the Marketing Authorisation and all marketing of the medicine must be consistent with the SPC.

INDEX

A

ABPI, 9
Accompanying persons
– attendance at meeting, 42
– payment for, 42, 51
Accurate, 15, 30
Activities – at a meeting, 45
Administrative charge, 13
Administrative staff – promotion to, 25
Adverse events, 55
Amusement park, 43
Approval procedure
– for promotional material, 12
Attendees at a meeting, 42
Audit, 35, 36

B

Balanced, 15, 30
BMA, 9
Brand name, 19, 22, 46
– of other company's product, 17, 31
Breach of Code, 13
– by representative, 14
Briefing – of external speaker, 49
Briefing material, 53
Budget holders, 29

C

Calls, 25
– arranging, 27
– convenience, 26
– frequency, 26
Call reports, 14
Call sheets, 14
Casino, 43
Certification, 12
– of representative-produced material, 21
Charity – donations to, 38
Chocolates, 33
Clinics, 36
Code of Practice Appeal Board, 13
Code of Practice for the Pharmaceutical Industry, 9
– breach, 13
– enforcement, 13
– scope, 10
Code of Practice Panel, 13
Code of Practice Review, 13
Company policy, 8
– **donations, 38**
– **hospitality costs, 44, 50**
– **meeting venues, 43**
– **representative letters, 22**
– **representative-produced material, 21**
– **sponsorship, 23, 37**

Comparisons, 17
Competitions, 34
Complaints, 13
– about representative activity, 14, 56
Consistency with SPC
– detailing, 30
– promotional material, 15
Contact details
– Adverse Events, 55
– Expenses, 56
– Medical Information, 54
Contract representative, 11
Cost per head, 50
CPD, 42

D

Data on file, 16
Delivery of a requested item, 28
Detailing, 29
Discounts, 39
Disguised promotion
– sponsorship as, 23, 37
Disparaging, 31
Distribution
– of giveaways, 33
– of promotional material, 20
Donations, 38
– as inducement, 28
– for holding a meeting, 51
– in lieu of hospitality, 50

E

Educational content of meeting, 41
E-mail, 20, 27, 29
Enquiries, 22
Exaggerated claim, 17, 30
Examination, ABPI Representatives', 53
Exhibition stand – see 'stand'
Expense forms, 14, 56
Expenses, 56

F

Fax, 20, 27, 29
Fee
– for interview, 27
– request for, 27
Football club, 43

G

General public, 25
Generic name, 19, 22, 46
Gift vouchers, 39
Gifts, 33
Giveaway, 33
– delivery, 28
GMC, 11, 27
Golf club, 43
Golf, 45
Grant, 37
Graphs, 18

H

Hanging comparison, 16
Health Authorities, 29
Health Authority Managers, 25
Health professional, 11, 13
– definition, 25
Highlighter, 20
Honorarium, 51
Hospitality – abroad, 52
Hospitality, 41, **44**
Hotel
– payment for, 50
– positioning stand at, 47

I

Inducement
– donation as, 28
– giveaway as, 34
– to gain an interview, 28
– to prescribe, 34
– sponsorship as, 37
Inexpensive, 33
Interactive data systems –
provision of P.I., 19
International conference, 29, 52
Invitation, 46
Invoice – for provision of hospitality, 50

L

Lecture notes, 46
Letters, 22
– to arrange an appointment, 27

M

MCA, 9, 13
Meals – payment for, 50
**Medical & Educational Goods
& Services, 35**
Medical Equipment, 35
Medical Information, 54
Meetings, 41
– **activities, 45**
– **attendees, 42**
– **hospitality, 44**
– **material, 46**
– organisers, 11
– **payments, 50**
– **presentations, 48**
– sponsorship, 51
– **venue, 43**
Misleading, 16, 30

N

New, 17
Nightclub, 43

O

Offence, 18
OTC medicine, 10

P

P. I. – see Prescribing Information
Package deals, 39
Patient confidentiality, 36
Patient data, 36
Patients – contact with, 36
Patients, 25
Payments – at a meeting, 50
PCO members, 25
PGEA approval, 41
Phone, 27, 29
Photocopies, 20, 24
PMCPA, 9
Post-it notes, 20
Practice managers, 25
Prescribing Information, 19, 22
– provision at a meeting, 48
– provision on meeting materials, 46
Presentation – at a meeting, 48
Primary Care Organisations, 29
Prizes, 34
Programme for a meeting, 46
Promotion – definition, 10

Promotional aid – see giveaway
Promotional material, 15
– distribution, 20
– use, 20
– approval, 12

Q

Quotes, 18, 31

R

Relevant to the practice of medicine, 33
Representative letters, 22
Representative-produced material, 21
– certification, 21
Reprints, 24
– available on stand, 47
– provision of P.I., 24
Restaurant, 43
Room rental, 50
RPSGB, 9, 11

S

Safe, 17, 30
Samples, 40
Screenings, 36
Signatory, 12
SPC, 19
– availability during detail, 32
– consistency with, 15, 30
– provision with sample, 40
Speaker (external), 11, **49**
– honorarium, 51

Sponsorship, 23, 37
— as disguised promotion, 23, 37
— as inducement to prescribe, 37
— of a meeting, 51
Sport — associated with a meeting, 45
Spouses — see 'accompanying persons'
Stand, 47
— provision of P.I., 19
Standard letters, 22
Storage — of samples, 40
Substantiation, 16, 31
Superlative, 16, 30

T

Telex, 20
Textbooks, 35
Third parties, 11
— letter produced by, 22
— material produced by, 21
Trade practices, 11, 39
Training material, 19, 53
Training, 53
— on Code, 12
Travel — payment for, 50
Travel log, 14

U

Undertaking, 13
Unlicensed product/indication, 29
— information about, 54
Unpublished data, 16

W

Wholesalers — hospitality for, 44
Wine, 33
Wine tasting, 45

CHECKLIST FOR MEETINGS

✓ HOSPITALITY

Does the meeting have a clear educational content and is education the main purpose of the meeting?	p41
Are all attendees either health professionals or appropriate administrative staff?	p42
Is the venue appropriate?	p43
Is the hospitality appropriate?	p44
Are any activities offered appropriate?	p45

✓ MATERIALS

Do all materials associated with the meeting comply with the Code?	p46 & 47

✓ SPEAKERS

Do all presentations, by you or by invited speakers, comply with the Code?	p48 & 49

✓ PAYMENTS

Are all payments appropriate and within company limits?	p50

✓ SPONSORSHIP

Is it clear that the meeting has been sponsored by your company?	p51

✓ AND FINALLY...

Is the overall impression of the meeting acceptable?